Archaeology: World Archaeology

An Introductory Guide to Archaeology

Miles Clarke

Copyright © 2015 Miles Clarke

All Rights Reserved

This is a work of fiction. Names, characters, places, and incidents are a product of the author's imagination. Any resemblance to actual persons, events, or locales is entirely coincidental.

This book or any portion thereof may not be reproduced or used in any manner whatsoever without the express written permission of the author except for the use of brief quotations in a book review.

The information provided in this book is meant to supplement, not replace, any proper training in the field of archaeology.

All images used in this book are Royalty Free and/or fall under the Creative Commons license.

ISBN-13: 978-1537249292

ISBN-10: 1537249290

Some Takeaways from This Book

Are you planning to take up Archaeology as a subject? Or did you just start studying archaeology to make a career? Either way you will need a brief of how the science works and what will be expected of you as an archaeologist.

Archaeology is that science or art – it can be maintained that is both – which is concerned with the material remains of the man's past. There are two aspects to the archaeologist's concern. The first of these is the discovery and reclamation of the ancient remains; this usually involves field excavation or at least surface collecting. The second concern is the analysis, interpretation and publication of the findings. To lay the mind an archaeologist is often a romantic figure that spends his time discovering hidden cities or royal tombs full of gold and precious stones. But the professional archaeologist spends most of his time in the classification and the interpretation of objects. The materials of archaeology are both the things made by man and things used by man. The things made by man are the settlements, buildings, utensils, tools, weapons, objects of ornament or pure artistic expression – the sum total of things fashioned in some way for human purposes. Strictly speaking archaeology is not concerned with the analysis and interpretation of the bones of ancient man himself – whether fossilized or not.

This e-book is informative and at the same time it will intrigue you to take up archaeology as a career. This book explains archaeology as a term, along with general history and development, the materials archeologists use to state spectacular findings and the future of archaeology.

Contents

Introduction 1
  What is Archaeology? 1

Chapter 1 - Brief History of Archaeology 7
  General History and Development 7
  The Mediterranean and the Near East................................... 10
  Inception of Archaeology 14
  Archaeological Developments of the 20th Century............. 18

Chapter 2 – The Materials of Archaeology 20
  Excavations and Field work 20
  Methods of Dating... 26
  Typology 28
  Cultural Inference.. 29

Chapter 3 – Underwater Archaeology 30

Chapter 4 – The Future of Archaeology 33

Chapter 5 – Famous Archaeological Sites............................... 35

Chapter 6 – Fun Facts about Archaeology 39

Page intentionally left blank.

Introduction

What is Archaeology?

The word archaeology comes from the Greek words *archaia* ("ancient things"), and *logos* ("theory or science"). The ancients themselves when digging holes in the ground for one reason or another often detected the remains of the pre-existent cultures and made investigations and derivations which might be called archaeology. Archaeology is that science or art—it can be maintained that it is both—which is concerned with the material remains of man's past. The first of these is the discovery and reclamation of the ancient remains; this usually involved field excavation or surface collection in the least. The second concern is the analysis, interpretation and publication of unusual findings. The archaeologist reconstructs the past activities and way of life – what they did for a living, the tools they used, the skills they deemed necessary for survival, and even the life-threatening epidemics.

The materials of archaeology include man's artifacts, from the very earliest stone tools, of perhaps millions of years ago, to the man-made objects that have become obsolete to man in the present day; everything made by human beings – from basic tools to state-of-the-art machines, from the earliest buildings for human habitation and chapels, mausoleums to castles, places of worship, and pyramids.

Introduction

The broken bones of fallen warriors in ancient times may reveal the weapons used in past warfare. Diseases leave a mark on bones; hence, they could be used to study epidemics of ancient times. The teeth would often reveal the diet of our predecessors. Animal bones can be studied for possible associations and use by ancient man. Excavation, in the present day, is a highly scientific procedure, though the voluntary help of amateurs seeking to make a career in the field is often welcome. It is not possible to dig up the entire site so the archaeologist plans trenches across it. The trench is dug up till they find soil untouched by man and each layer of soil is used by the archaeologist to draft an initial document of premature findings. Anything of importance redeemed of the site is most important to culminate the findings.

As generally believed, archaeology is not the study or interpretation of the bones or human fossils of ancient man. The study of bones is restricted to physical anthropologists or human paleontologists. Neither is an archaeologist supposed to elucidate ancient writing since that is the specialty of an epigraphist or philologist. Ultimately, the archaeologist is a historian and primarily an anecdote: he has to discover, explicate, classify and evaluate the artifacts that are to be studied. Many good archaeologists spend their entire lives in the activity of epitomizing and classification of the material remains in the ancient contexts, to codicil the existing knowledge, and, thus to augment the understanding of the past.

The archaeologists use scientific methods, techniques and even expertise to fully decipher facts related to an ancient discovery. The ancient relics must often be studied in their natural environmental contexts; and zoologists, soil scientists, geologists, botanists, anthropologists and other experts are often invited to determine and diagnose plants, animals, soils, human remains and rocks. Archaeology is not a natural science, despite the excessive use of scientific methods of the physical and biological sciences. Some consider it as a discipline that is part science and part humanity. Perhaps, an archaeologist is rather an artisan, specialized in his crafts, rather than a historian.

The extenuation of an archaeologists work is in the substantiation of all history, so to ameliorate the present by our knowledge of the experiences and intellectual or physical conquests of our predecessors. The most direct findings of archaeology lug on the history of art and technology because it largely consists findings of the ancient man-made things. But by conjecture, it also yields knowledge of religion, society, economy and a complete way of life. Also, it may bring to light and decipher previously unexplored written documents, providing an even more certain form of evidence about the past.

Introduction

Considering the vast nature of archaeology as a discipline it is not possible for a single archaeologist to study the entire human race's history. There are many offshoots to archaeology that categorize it by geography and period. Under the subsection of geographical or classical archaeology, the study derives its interest from the previous Greek and Roman civilizations. Archaeology as a discipline finds its roots within the interest of discovering the way of life of the Greek and Roman civilizations. This is where classic archaeology found its origins as a vast segment of the study. Egyptian archaeology originated with Napoleon's invasion in 1798. He brought with him scholars who took a keen interest in maintaining an archaeological record of the remains of the country. They were able to elucidate the numerous Egyptian writings which further enabled scholars to understand Egyptian history and culture in a manner unlike any other.

On the other hand, archaeology by periods is an extension to classical archaeology and it includes the study of medieval times and the industrial revolution. The initial findings of ancient writings fall under prehistoric archaeology, which hails back to some 5,000 years in Mesopotamia and Egypt with some existence in India and China and later in Europe.

Prehistoric archaeology goes back to 500,000 years when human development, climate and the environment was constantly evolving. But it's since the middle of the 19th century that it has been referred to as prehistoric and, it's under this subdivision that the archaeologists solely have material, and environmental sources to draw conclusions and transmute the facts to history.

Archaeology has immense potential due to the incompleteness of the archaeological record. An imperforate archaeological record is possible if each and every historical relic, and all non-artifact material, used by man since the beginning was already redeemed in its natural state and was ready for annotation. Such completeness of record indicates more archeologists are needed to examine the situation. It is the association of the artifacts and non-artifact material recovered in its original state; how the ancient man used it and the complete physical manifestation of a way of life that are the true basis for a complete archaeological record. Although, it is evident that each and every confluent relic or artifact will have a lower survival value. But archaeology derives it techniques from science to recover and preserve perishable items.

Introduction

The scope of this book is to describe briefly the context of archaeology, how it came into existence as a learned discipline; how the archeologist works in the field, environment, laboratory, and how he analyzes and deciphers his material evidence and transmogrifies it into history.

Chapter 1 - Brief History of Archaeology

General History and Development

The earliest provenance of Archaeology was in Europe during the 15th and the 16th century era when the Ionics of Renaissance cast a look upon the eminence of Greece and Rome. During the 16th Century, the Pope and cardinals in Italy began to amass relics of the past and widely spent on excavations to discover ancient art. These patrons of ancient remnants were copied by their counterparts in Northern Europe. Up to the 18th century, artifacts were considered to be toys of the witches and dwarfs, only a few well-bred minds such as of Leonardo da Vinci could have comprehended the importance of such remnants. At this point, it is clear that one of the two major approaches to archaeological reasoning arose with Renaissance. Sculpture, pottery and coins were collected, and exhuming of artifacts was conducted, primarily to sell to the richest collector. During the Revolution in France, museums were erected and the most prestigious Louvre was inaugurated – a need for historical artifacts arose to the display halls. However, all this activity was still not archaeology and was rather art collection.

Chapter 1 Brief History of Archaeology

At this point, the humanistic approach to archaeology can be divided into two branches: the fine arts or museum tributary and the academic or scholarly division. Given the philosophical bearings of the humanistic approach to archaeology has tended to constrict itself to Greece. With tenuous exceptions, the humanistic approach has concerned itself only with the earlier instances of the western culture. Some parts of Europe, France, Great Britain and Germany were all part of these exceptions for their interest in historical artifacts.

The second approach to archaeology took shape during the 19th Century and it was popularly called the approach of the Social Sciences. This was more of the early origins of anthropology. This indoctrination of anthropology was pushed by the theories of Charles Darwin, Thomas Henry, and other great geologists. However, they gave rise to an approach that the humanists had no interest in whatsoever, because the 'alluring', the 'incidental', the very far-flung people, whether dead or alive were of no interest to the humanists. Despite the difference, anthropological archaeology developed museums, exhibitions and other scholarly branches to the approach. As sovereign approaches to archaeology, both were inimical to the common goal that is to fully understand mankind.

Both the approaches gained denotation but each developed a professional pompousness to itself. Archaeology's major concern was to understand mankind and the past and the only means of gaining this knowledge was exploring extinct human cultures and achievements that were penned down by ancestors.

Approximately 500,000 years of prehistoric human development the climates and the environment changed and of course, there were no boundaries, hence, it is difficult to gather whether the inhabitants of France were actually the Frenchmen back then, or the Britons actually occupied their land some hundred years ago. Paleolithic archaeology is concerned with the inception and augmentation of early human culture, which is believed have occurred some 600,000 or 700,000 years ago. Although there is no way to prove but modern evidence suggests that the earliest human forms had deviated from the familial lemur stock by the genesis of Pleistocene. It is impossible to actually prove the existence of early human forms but archaeology as a discipline uses scientific techniques to gather information from fossils or remnants discovered during excavations at historical sites.

Chapter 1 Brief History of Archaeology

THE MEDITERRANEAN AND THE NEAR EAST

Archaeology as a discipline took shape during the 18th Century, with the sudden and profound significance of the Greek and Roman civilizations. In this area, especially in Israel, Jordan, Lebanon and Syria, a Lower Paleolithic development closely paralleling that of Europe is indicated by the widespread distribution of hand axes of Abecvillian and Acheulian type. Unfortunately, the majority of these finds are from open-air, unstratified sites that cannot be dated. A crude flake industry, reminiscent of the Tayacian of Western Europe, has been reported from several cave sites.

During that era, widespread excavations subsisted in ancient Italy including the cities of Herculaneum and Pompeii. Heinrich Schliemann work on ionic archaeology entrenched the discipline on a scientific basis, and he probed the provenience of Greek civilization particularly in Troy and Mycenae (1870's). Similarly, M.A. Billotti, at Rhodes during the same period; of the German Archaeological Institute, studied the Greek civilization. During 1873 to 1875, Alexandar Conze at Samothrace was the first one to add photographs in his book to support the study. Schliemann, during the same period affianced to dig the Crete, but when he could not do it, Arthur Evans in 1990 instead undertook the task of exploring Knossos in order to unearth the Minoan civilizations who were the progenitors of the classical Greece.

The scholars who accompanied Napoleon to Egypt began the modern archaeological study of that country, although hampered by the fact that they did not understand the ancient language. The outcome of their work was published in the *Description de l'Egypte* (1808-25). Historic Egypt emerged in the first dynasty with a developed hieroglyphic writing and a brilliant display of technical skill. The magnificent objects excavated by Walter B. Emery and Zaki Saad in the Sakkara and Helwan cemeteries of the Memphis district lend a new meaning to the fragmentary material known earlier from the shrine at Hierakonpolis and the Abydos royal cemetery in upper Egypt.

When Jean Francois Champollion succeeded in deciphering the hieroglyphic text of the Rosetta stone in 1821-22, rapid advances were soon made in the translation of Egyptian texts. This was the pristine footprint in Egyptian Archaeology as a discipline and this effort to read and understand the many great ancient writings of Egyptians was the beginning of an era; as a result many scholars followed him on this path of ancient discovery. Antiques and relics from Egypt rose in value during this time, due to an increasing demand to own such prestigious artifacts. This also led to methodized pillaging of the Egyptian tombs by rebellious raiders. The 19th Century also saw the formation of collections of antiquities and the first accurate copying of the monuments.

Chapter 1 Brief History of Archaeology

Some excavations were undertaken by Auguste Mariette, who found the antiquities service and the Cairo museum. However, scientific excavation was a development of the 20th century, based on methods devised by Sir Flinders Petrie and developed by such men as Ludwig Borchardt, George Andre Reisner and Herbert Winlock. In conjunction with this field work, immense contributions were made by the philologists Adolf Erman, Kurt Sete, F.L. Griffith and Sir Alan Gardiner and by the historian Gaston Maspero, Eduard Meyer and James Henry Breasted. Petrie began excavating in Egypt during 1880 unearthed marvelous artifacts and cultural findings there; later he did outstanding archaeological work in Palestine. Petrie developed a systematized method of excavation, the fundamentals of which he summarized in his book; *Methods and Aims in Archaeology* (1904). Hereon, Howard Carter and Lord Carnarvon made the most prodigious discovery in the history of Archaeology, that of the tomb of Tutankhamen in 1922.

In contemporary archaeological usage, the name Mesopotamia is generally applied to the region of Western Asia which was dominated by the river valleys of the Tigris and the Euphrates. The Old Testament was, indeed, for many centuries the main source of information about ancient Mesopotamia available to the western, European, world and the main stimulant for interest in that region.

During the middle ages, few Europeans ventured into this relatively remote, inaccessible and inhospitable region. The era of regular archaeological excavations was opened in 1842 by the French consul in Mosul, Paul Emile Botta, who began his activities at Nineveh but soon transferred them to Khorsabad, where he discovered and began the exaction of the palace of Sargon II (722-705 BC). The British followed almost at once with excavations at Nineveh and Nimrud under the direction of Sir A.H. Layard. The results of both enterprises far exceeded all expectations. During the first decade of regular excavations, the French and the British discovered palaces of Assyrian Kings, which were decorated with thousands of bas-reliefs depicting the daily life of the Assyrian court, with colored glazed bricks and paintings.

In 1846, Henry Creswicke Rawlinson became the first person to expound Mesopotamian writing called cuneiform. During the last quarter of the 19th century, methodical excavations led to the espial of human cultures and people previously unknown, such as the Sumerians, who were the predecessors of Babylonians and Assyrians in Mesopotamia. Leonard Woolley made the most spectacular Sumerian excavation in 1926, of the Royal Tombs at Ur.

Chapter 1 Brief History of Archaeology

INCEPTION OF ARCHAEOLOGY

Scientific archaeology matured during the 19th century in Europe from the paleontology and treasure hoarding of the 16th, 17th and 18th centuries. This transpired due to three major reasons: Revolution of geology, an insurgency of antiquarianism, and the proliferation of the theory of evolution. Geology saw its primate revolution during the 19th century with the exploration of and manifestation of uniformitarian stratigraphy – which impels the age of the fossil remains by the gradation they occupy under the earth. Geologists like William Smith, Georges Cuvier, and Charles Lyell made these discoveries possible hereunto, making it a discipline. Lyell in his book Principles of Geology (1830-33), familiarized this new system and led to the acknowledgement of the great relic of humankind.

Lyell's principles were regarded well by Charles Darwin as one of the burgeoning ground works to the formation of his theory of evolution. The 18th Century saw the discovery of Man's first tools that were ever used in history and the location of these historical relics at a certain level manifested from the principles of uniformitarianism, that proved man's existence long before 4000 BC, which is the date mentioned in the Bible.

Jacques Boucher de Perthes discoveries in the Somme valley of France, and also William Pengelly's in the caves of South Devon- England; were used to exhibit the relics of man's history in 1859, and this was the same year, Darwin's revolutionary publication *Origin of Species* was published. The Paleolithic Period of man's history was established during this era, although the term 'Paleolithic' was first used by John Lubbock in his book *Prehistoric Times* (1865) and it was after this that the expression began to be used widely by archaeologists, geologists, historians and anthropologists world-over.

Scandinavian archaeologists created a debacle in antiquarian thought by hypothesizing, in the sphere of archaeology, the sequential technological stages in man's history. The Copenhagen museum which opened to the public in 1819 was materialized by C.J Thomsen on the basis of three consecutive stages of Stone, Bronze, and Iron. J.J.A Worsaae, who was a pupil of Thomsen, proved that the placement of the stages in the museum was accurate with stratigraphy in the Danish quagmires and mounds. During the mid-1850's, the low lake levels in Switzerland led to the excavation of the primitive Swiss lake habitation, and, here again, the theory of sequential technological stages of man was confirmed.

Chapter 1 Brief History of Archaeology

Darwin's Origin of Species connoted a long history of mankind, and also the compliance of human evolution as a true theory. This acceptance created a climate of change during the last four decades of the 19th century, archaeology as a discipline flourished most during this period and which further led to augmentation in the unraveling of the complete story of man's development. Lubbock, during the prehistoric times, elaborated the three-age system of Thomsen and Worsaae to a four-age system, hence dissecting the Old Stone Age concept into old and New, Paleolithic and Neolithic, respectively.

The last quarter of the 19th Century saw remarkable Paleolithic discoveries made in Spain and France; these included different forms of art namely, sculpture and cave paintings from 30,000 – 10,000 BC, which is now known as the Paleolithic period. Marcellino de Sautauola first discovered the cave paintings in Spain during 1875-80; they were disregarded and not accepted as Paleolithic paintings. It was when similar paintings were discovered in Les Eyzies in France around 1900, that those paintings were accepted as the first discoveries of its kind and also the most spectacular archaeological findings in history. Similar findings were added to the Paleolithic successions of archaeological discoveries during the 20th Century. The most prominent ones were the discoveries at Lascaux, France, in 1940.

The last quarter of the 19th Century saw the primary groundwork laid for scientific archaeology with the amazing discoveries that were made with Gen. A.H. Pitt-Rivers's excavations of prehistoric and Roman archaeological sites at Cranborne Chase, Dorset. Scientific archaeology was improvised as a technique and re-used in England and Wales by Sir Mortimer Wheeler and Sir Cyril Fox.

Chapter 1 Brief History of Archaeology

ARCHAEOLOGICAL DEVELOPMENTS OF THE 20TH CENTURY

Archaeology was defined as a discipline during the 20th Century and it was extended further from the old favorite archaeological sites; Near East, the Mediterranean, and Europe to the rest of the world. Excavations at Mohenjo-Daro and Harappa, in present Pakistan, proclaimed the existence of the Indus civilization. In Eastern China, An-yang unleashed the existence of prehistoric Shang dynasty which is proof of ancient Chinese culture.

The Stone Age was continued to be studied by archaeologists during the 20th Century and more astounding discoveries were made which includes, revelations by L.S.B. Leakey, of stone tools and skeletal remains of early man dating back 2,000,000 years in the Olduvai Gorge in Tanzania. Some Neolithic archaeological sites were exhumed at Jericho in Palestine; Hassuna, Iraq; Catalhuyuk, Turkey; and other parts of the Near East. These findings established the inception of agriculture.

Archaeology took speed in America much later than it did in Europe. The earliest discoveries in America were during 1784, by Thomas Jefferson who excavated peats and barrow in Virginia. Prehistoric America, was discovered to great extents during the 20th Century, and the most spectacular discoveries was that of the domesticated crops, Maize in Central America, and the Olmec civilization in Mexico (1000 – 300 BC). This is by far the oldest civilizations to have walked the earth. During the 20th Century, archaeology developed as an academic discipline, and most of the well-established universities around the world have professors and a department of archaeology. There have been studies, research, books and journals in this field of study that aim at helping everyone understand archaeology as a subject.

Chapter 1 Brief History of Archaeology

Chapter 2 – The Materials of Archaeology

Excavations and Field work

Archaeology is the science that studies the history of human culture through its material remains, is limited to analysis of those items (and inferences drawn there from) which have escaped the destructive forces of decay. Wherever man has altered the natural surface of the earth, as at a site where people have lived, some evidence will remain, and with the proper method of detection, such spots can be identified. Usually, the archaeologist concentrates his field activity at a spot where intensive or long-term human activity has occurred, for there is the location most productive of the information for which he/she is searching.

Archaeological sites may be of several sorts. Closed sites such as caves may on occasion be so well protected from the weather that otherwise perishable objects made of wood, basketry, or leather may be perfectly preserved. In the western United States, for example, there are a number of dry caves which have produced large quantities of well-preserved organic materials, and the complete preservation of some tomb contents of dynastic Egypt are well-known. In northwestern Europe a large number of remains, either of individual objects or burials or of whole sites, have been found in peat deposits.

Organic materials tend to be preserved nearly as effectively in the moist, acid conditions of peat as in completely dry caves. Most prehistoric sites are in the open, and consequently are subject to the influences of climatic and geologic forces which cause the disintegration, covering or erosion of evidence from former occupations. Trash mounds, formed by the gradual accumulation of refuse, occur very generally where man has settled for any length of time. In coastal regions such sites commonly take the form of shell mounds, so named because their bulk is largely composed of shells of mollusks once used for food. Fire hearths, floors of houses, storage pits, lost and discarded tools, bones of food animals and all the miscellaneous by-products of living commonly occur in such trash heaps. Cemeteries where the dead were buried usually occur in the proximity of occupation sites.

Chapter 2 The Materials of Archaeology

Megalithic or other stone or earth structures such as temples, dolmens, barrows, fortifications, ditches, roads, and pyramids may be particularly obvious remains of man's former presence. Other sites can be in the form of cave walls bearing pecked petroglyphs or painted petroglyphs such as the famous Upper Paleolithic caves at Altamira, Spain, and Lascaux, France. This general type of aesthetic expression occurs widely in all continents. Mines and quarries where flint or obsidian, soapstone, building stone, colored pigments, iron or copper ore etc were extracted which are evidence of early industrial pursuits.

Even ancient garden plots can be located, as proved by aerial surveys in the British Isles. In addition to material objects made by man, food remains in site deposits may enable a reconstruction of economic pursuits and provide some indications of the relative importance of fishing, hunting, gathering or farming. Trade, travel and intergroup contacts can often be reconstructed by identifying the material or form of certain objects as made in some region far distant from the spot where they are recovered by an archaeologist. The aim of the field archaeologist is to recover as much material as possible or necessary for the interpretation or reconstruction of the former culture pattern. Since archaeology means systematic excavation for the recovery of buried materials, the excavator must be acquainted with the techniques of map-making, methods of preserving materials which he uncovers and correct procedure in actual digging.

Discovery of archaeological remains may come about through several means. Materials may be exposed through natural erosion, as in the banks of rivers or streams, or by human disturbance, as in ditching, plowing, or digging or building foundations.

Chapter 2 The Materials of Archaeology

Commonly the field archaeologist is an expert in recognizing surface evidences of man's former presence in the form of a distinctive dark soil color due to charcoal from fired, irregularities in the normal surface relief, unusual vegetation which is a function of the different soil chemistry of occupation sites, or the presence of exposed artifacts as a result of erosion of the surface. Once having searched for and found the sites, the archaeologist selects the site to be excavated.

Excavation entails careful attention to the placement and association of each object or feature uncovered. Most or all portable objects are saved and removed to the laboratory or museum for study. Non-portable remains such as house floors, stone constructions are recorded in notes and documented in photographs.

Bones of animals are collected and identified and this information may indicate foods used as well as give evidence of the local environmental conditions when the culture was in operation. The association of items in the deposit may assist in determining their original functions; thus, objects found in a ceremonial chamber, in a refuse pit or in a cache of tools may indicate, respectively, that the various items were used for religious, culinary or industrial activities. Full photographic and notebook records of finds and observations are made, and these are utilized in preparing the final report.

Some excavations, especially in the Mediterranean and Central American areas, were a monumental architecture which is notably characteristic, and it can be performed primarily for the purpose of reconstructing ancient temples or pyramids. In every case, the particular problem which the archaeologist is attempting to solve determines his choice of sites and method of approach in an excavation.

Chapter 2 The Materials of Archaeology

METHODS OF DATING

Since Archaeology is a historical discipline and concerns the development and succession of human cultures, the problem of dating finds is of fundamental importance. Age assignments may be of two orders: direct or absolute dates; and indirect or relative dates. Direct or absolute dates can be expressed in numbers of years. Calendar records in the old world; do not carry beyond the last 5,000 years and in the new world not past the 21st Century. The absolute dating method known as dendrochronology works by counting annual rings of trees. Beginning with living trees, the pattern of wide and narrow rings, which reflect conditions of temperature and rainfall experienced during the life of the tree, can be extended backward in time with the use of prehistoric wood samples.

This method was developed and carried to its highest point in the southwestern United States, where the tree-ring chronology extends back about 2,000 years. Elsewhere, as in Alaska and the eastern United States, the method has been applied to advantage but without such striking results. Paleolithic implements and fossil human remains which occur in identifiable glacial or interglacial deposits can thus be dated approximately with reference to the Zeuner chronology.

An alternative theory of causes for Pleistocene glaciations has been proposed by M. Ewing and W.L. Donn. Counts of annual sediment layers deposited by retreating glaciers in northwestern Europe and northeastern North America have yielded a chronology, to which archaeological finds can often be correlated, for the last 15,000 years. Another method of absolute dating is the radiocarbon technique. Mildly radioactive carbon is formed in the atmosphere and is contained in the bodies of all organisms. C14 decays or disintegrates at a constant rate, its half-life being 5,700 years. Relative or indirect dates are expressed relative to some other chronological event.

The Pleistocene glacial chronology and a paleontological succession of animal forms furnish convenient consequences of events to which Paleolithic culture finds may be referred for relative age determination. The primary chronological tool of the archaeologist is stratigraphy or superposition. Unless disturbances of deposits have occurred, the oldest objects will lie near the bottom and the most recent ones near the top. Analysis of stratigraphic series of pollens contained in postglacial lakes of bogs has yielded surprisingly complete information on the sequence of ancient climates and flora which can often be of aid in assigning age's prehistoric archaeological materials. Another method of deriving chronology or sequence from surface materials where stratification is lacking is called seriation.

Chapter 2 The Materials of Archaeology

TYPOLOGY

While much of the information derived from excavations can be presented by simple description and illustrated by drawings or photographs, this is not always convenient, since there may be large numbers of specimens which show important individual differences. The usual means of treating large bodies of artifacts is by means of typology or classification into classes and types. Classes consist of groups of objects segregated according to material and further segregated into subclasses or types based on differences of form, function, decoration or technique of manufacture.

Classification may be based upon the methods of manufacture, as in the case of Paleolithic stone implements, where shape may have been incidental and the flaking process was the primary concern of the maker. By mapping the location of sites and determining the geographic distribution of cultures it is often possible to find correlations of culture areas with natural environmental areas, to determine settlement patterns and shifts of population centers through time, and by comparing the local or regional culture sequences to gain insight into culture-historical processes over wider areas.

CULTURAL INFERENCE

Once having performed the labor of excavation and the task of classification, it remains for the archaeologist to treat the historical problem of how the culture he is studying came to constitute the particular aggregation of traits which it displays. Such explanations, which are rarely complete and precise, entail reference to what are called culture processes, the factors operating toward the change, growth or stabilization of cultures.

Culture types which are geographically widespread may be assumed to have diffused from one generative center in the course of time, and the point of origin can often be indicated as the locus where it occurs earliest or in most complex form. Proofs of such alternative explanations can rarely be adduced, and the decision to support one or the other usually rests upon that one on which the student believes the weight of probability to bear most heavily.

Chapter 2 The Materials of Archaeology

Chapter 3 – Underwater Archaeology

Underwater archaeology involves, for the most part, short-term perspectives because migrations within the open Pacific could have occurred only after the development seagoing canoe navigation in Neolithic times. The exception is the New Guinea, Australia regions, where the ancestors of the Australoid and Negritoid-types evidently arrived in the Paleolithic times. Records of early travelers described some of the most spectacular sites e.g. the giant statues of Easter Island and the Royal tombs of Tonga.

By the late 19th century, a few scientific institutions and private scholars were carrying on sporadic archaeological work. Occasional finds were made in caves, as in Hawaii, or by rough excavation as when F.W. Christian in 1896 tells of tearing out the floors of tombs in the so-called Venice of Ponape. Otherwise, finds were almost all surface materials; tools, stone constructions and petroglyphs. The long-term history of the oceanic people, especially the Polynesians, has been the subject of many speculative theories.

Professional scholars reject ideas involving a lost continent or direct with the inner east India. They also insist that while eastern-voyaging Polynesians could well have reached the American continent, and some may have found their way back into the islands, none of the various theories which claim that oceanic people had their homelands in North or South America are scientifically credible. Similarly, they reject theories explaining the pre-Columbia civilizations on the American content in terms of influences by way of the tropical Pacific islands from Asia.

The archeological record begins when early Homo sapiens populations, comparable with the fossils of Wadjak in Java, Aitape in New Guinea and Keilor and others in Australia, were moving eastward. This apparently occurred during the fourth glacial epoch when sea levels were lower, land pathways perhaps generally more uplifted and interisland channels narrower than now. Early man could then migrate with lessened water obstructions from the Asiatic continental platform through the intermediate Celebes-Molucca-Lesser Sunda zones on to the Australian continental platform. Some scholars suggest such movements during the third glacial, but this seems very dubious. By contrast, archaeological work in Polynesia and some zones of Micronesia is considerably more advanced. For northwest Micronesia, a vital clue is a radiocarbon dating of approximately 1527 B.C from a stratified deposit excavated by A. Spoehr in the Marianas.

Chapter 3 Underwater Archaeology

Spoehr estimated from the collateral evidence that occupation may go back to 2000 B.C. Spoehr suggested that for archaeological purposes Micronesia and Polynesia may usefully be treated as a continuous zone.

Such local constructions, however, together with other more spectacular elements such as widely scattered petroglyphs and a dubiously old "script" on Easter Island, have less significance for historical reconstruction than detailed study of variability in minor artifacts. Theories of contact with the American continent must be treated with caution so far as they lean on gross parallels such as stone images or art resemblances; the most concrete evidence by the 1950s was the presence of the sweet potato, apparently an American plant.

Chapter 4 – The Future of Archaeology

Archaeology is now a discipline and a field of study with societies and scholars actively working within the realm of the subject. The first archaeological society met in London around 1585 or 1586. An outgrowth of the antiquarian interest of the 16th Century, it was formed by a group interested in collecting and preserving British antiquities. During the 18th Century and early 19th Century increased interest in local archaeological remains led to the formation of other similar societies. Archaeology has always exhibited an upward trend when it comes to interest in the particular field of study or work. Archaeology is an evolving discipline and it has been often called the "handmaiden of history."

This interest in national history and culture has grown with time, and attention has been directed to classical antiquities of the World today and also the past. A number of societies have erupted that financially support excavations at new sites each year. There are about 22,000 members of the national archaeological organizations in the United States alone, but archaeological programs in the academia are a handful. Archaeology as a discipline may never fit easily in academic or a departmental nice at traditional universities of America. But it is however, the study of humans and human behavior. It explores the relationships between humans and their environment, based on material remains from the past.

Chapter 4 The Future of Archaeology

Archaeology covers natural and social sciences as well humanities, hence it is a vast discipline. Archaeological science is expensive and it does not attract a lot of research funding, but small policy adjustments at the State level can help encourage the practice of such a science. The face of archaeology has changed with time since there are computer based tools to classify and interpret objects and 3-D layers of information. The advancement of tools, techniques and methods to practice Archaeology make it the most interesting discipline today. Archaeology is and will always be a good career choice since one gets the fame and appreciation along with monetary benefits. The future of Archaeology is bright and it will evolve into a more advanced way studying man's past.

Chapter 5 – Famous Archaeological Sites

Some of the Archaeological sites discovered in the world are not just intense but also awe-inspiring. Here a few intriguing discoveries of the past:

- ✓ **Roman Bathhouse:** This was an intense finding since it revealed a lot about the lives of the Romans and the Byzantines but, what's more, compelling is that the bones of hundreds of babies were found in this bathhouse. It could have been a sacrifice or anything else.

- ✓ **Aztecs:** Aztecs were known for their sacrifices and they hosted sacrificial festivals where one grim discovery was made of outside Mexico City. A number of mutilated human bodies and animals were found stacked up.

- ✓ **Machu Picchu:** The famous ancient Inca citadel in the Andes, which was discovered in 1911. Machu Picchu is currently recognized as one of the New Seven Wonders of the World.

- ✓ **Terra Cotta Army:** The Terra cotta army was buried with Qin Shi Huang, the first emperor of China and the intention was for the soldiers to protect their emperor in life after death. Talk about sordid belief systems?

- ✓ **Pyramids & the screaming mummies:** The pyramids are a significant archaeological site but often Egyptians ignored the fact that strapping the chin to the skull will lead to the mouth falling open in a permanent scream. Although, many mummies were found screaming at the time of death due to some ritual torture.

- ✓ **Chemical warfare:** During an excavation at the ruins of the Roman/Persian battlefields, Robert du Mesnil found tunnels that headed to the city and inside the tunnels were remains of Roman soldiers trying to run while Persian counterparts holding them back. This was a trap and as soon as the Roman soldiers fell through they encountered burning sulfur and bitumen which turns into acid inside the lungs.

- ✓ **Egyptian hieroglyphs:** Egyptian writing was one of the most spectacular discoveries in archaeology. These were carved on a stone around 200 BC.

- ✓ **Troy:** This is one of the major archaeological discoveries and given its history the city is a legend. Situated in Turkey, this has been one of the most favorite areas for excavation over the years.

- ✓ **Easter Island:** This is one of the most isolated places in the South Pacific, but it is a remarkable archaeological discovery. It is interesting that humans managed to settle in this area but also managed to construct enormous stone heads around the island.

- ✓ **Dead Sea scrolls:** The Sea scrolls are one of the oldest surviving copies of the biblical documents dated way back to 150 BC and they are one of the major archaeological finds in history.

Chapter 5 Famous Archaeological Sites

These are just a few of the spectacular archaeological finds from around the world, there thousands of more discoveries and even more that are yet to be discovered by the future archaeologists. 'A few years down the road, that archeologist could be you!'

Chapter 6 – Fun Facts about Archaeology

- ✓ Did you know archaeologists **do not** dig up dinosaurs as it is widely perceived to be? They study human history.

- ✓ Archaeologists get to travel **all over the world** and all the time. Talk about being homesick?

- ✓ Archaeologists **do not** get to keep any of the things they discover. It goes to the museum or the laboratory.

- ✓ Did you know Archaeology is also called the "handmaiden of history?"

- ✓ Archeology lets grownups play in the dirt.

- ✓ You know a popular archaeologist from the movies – It is Indiana Jones!

- ✓ Archaeologists study things as intense as the pyramids and also things as small as the pollen spores.

Chapter 6 Fun Facts about Archaeology

Thank you for reading this book, I hope you enjoyed it. Please see below for more interesting books in my catalog, to expand your knowledge in world sciences.

Happy reading!

-Miles Clarke

Astronomy for Beginners: The Magical Science of Stars, Galaxies, Planets, Black Holes, Wormholes and much, much more!

www.ingramcontent.com/pod-product-compliance
Lightning Source LLC
Chambersburg PA
CBHW070415190526
45169CB00003B/1272